AF266696

# 4th Grade Math
# Volume 1

---

---

ISGN: 978-1-939796-82-0

# Table of Contents

# Place Value

**Key Vocabulary**

ten thousand

hundred thousand

million

ten million

hundred million

**What is the value of 7?**
Write your answer next to the number.

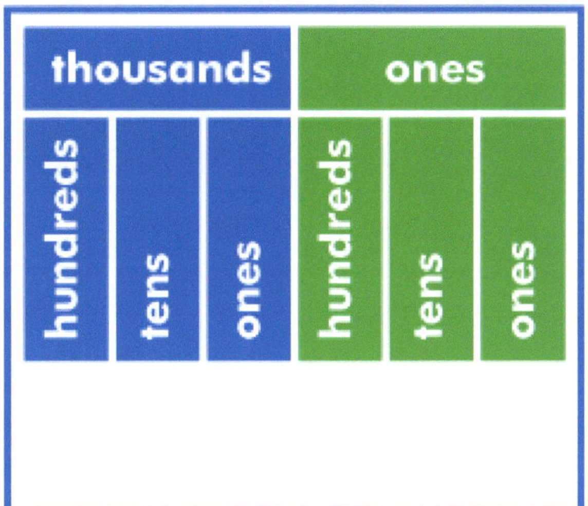

2 **7** 5,4 3 8

2 **7**,2 0 2

3,**7** 0 4

**The Value of Each Digit**

| thousands | | | ones | | |
|---|---|---|---|---|---|
| hundreds | tens | ones | hundreds | tens | ones |

$$2\ 5\ 6,1\ 4\ 7$$

$$200,000$$
$$50,000$$
$$6,000$$
$$100$$
$$40$$
$$7$$

**What is the value of 2 in each number?**

52,998

234,003

9,288

28,707

**The Number After 999,999**

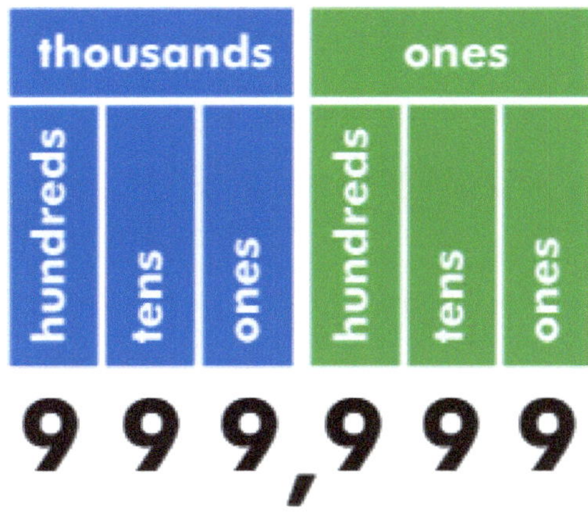

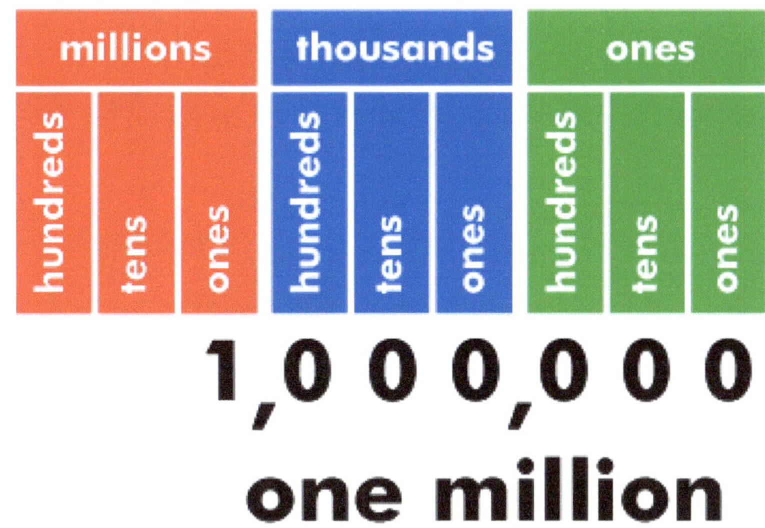

**State Population Data**

What state's population has a 2 in the hundred thousands place?

What state's population has a 5 in the millions place?

What state's population is more than 5 million?

**Write these amounts in the place value table.**

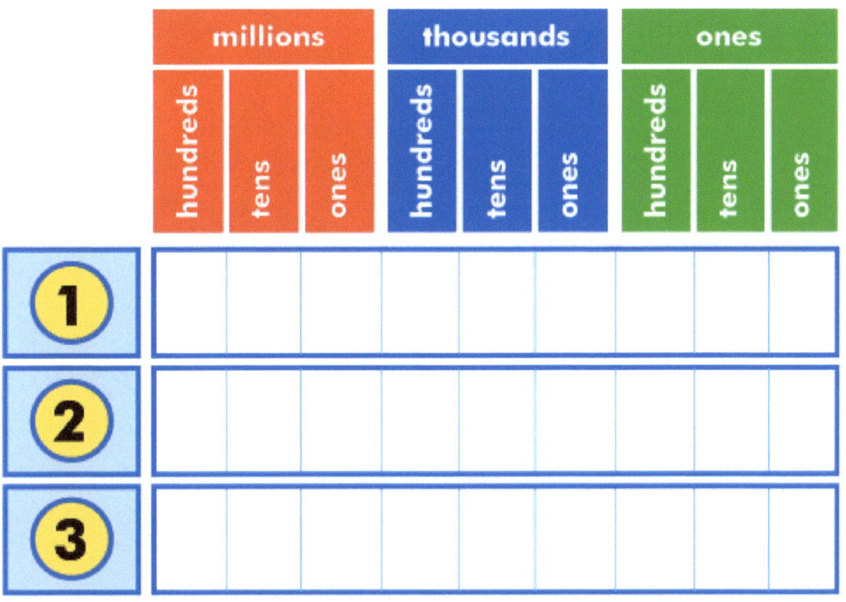



**1** Two million four hundred thousand

**2** Twelve million three hundred five thousand

**3** One hundred five million six thousand four hundred

Name________________________________

## Place Value Quiz

Circle or fill in the correct answer.

**1** **True or false? The value of the red digit is 8 million.**
**6,845,098**

**2** **What is the value of the red digit.    25,237,900**

**A** fifty million

**B** five million

**C** five hundred thousand

**D** fifty thousand

**3** **Write forty-five thousand nine hundred**

**4** **Write eighty thousand eight**

# Addition & Subtraction

**Key Vocabulary**

sum

difference

regrouping

ones

tens

hundreds

**Regroup using the fewest number of 10 base blocks.**
Draw your answer in the boxes provided.

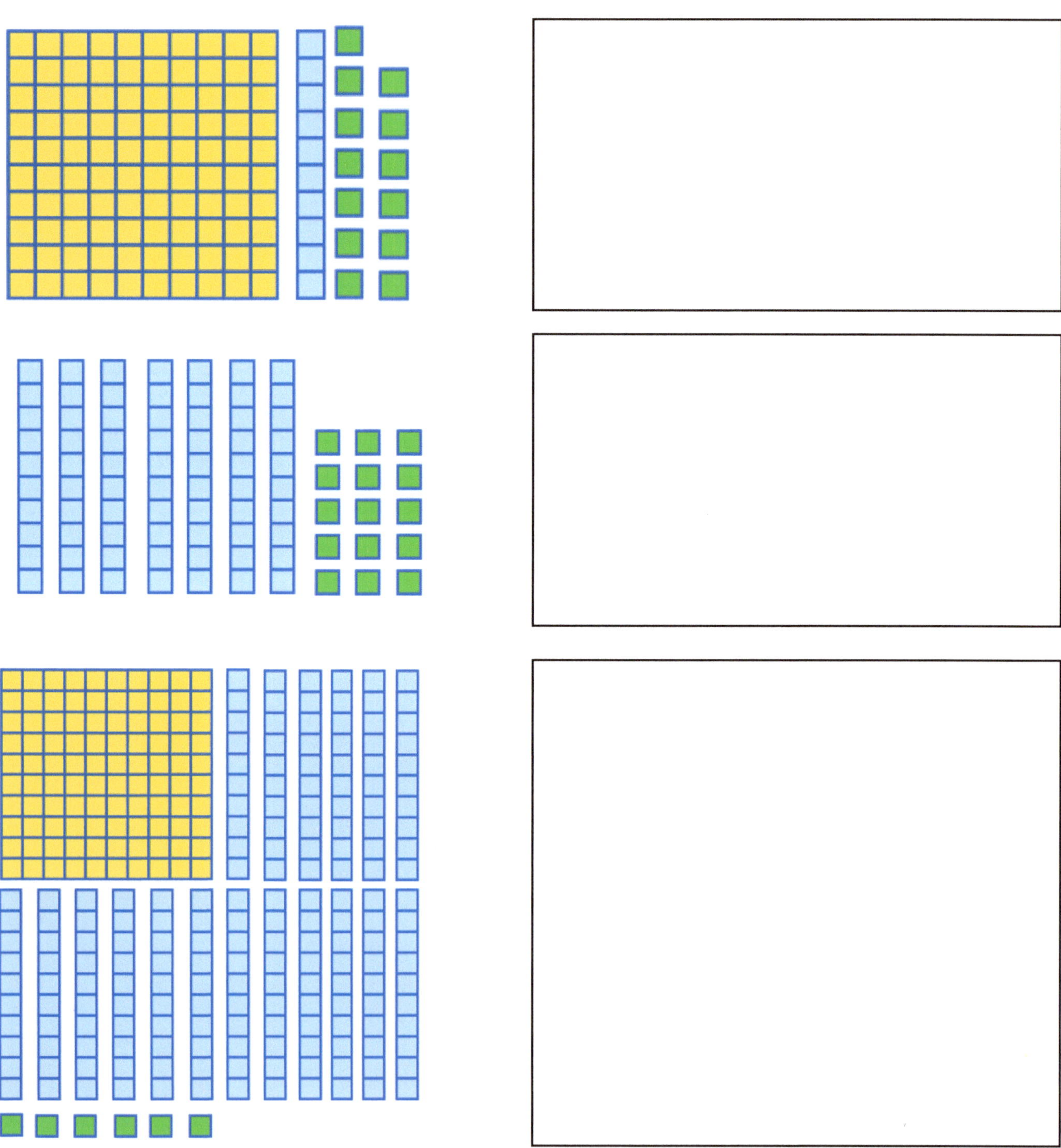

**Model the solution to this addition problem.**

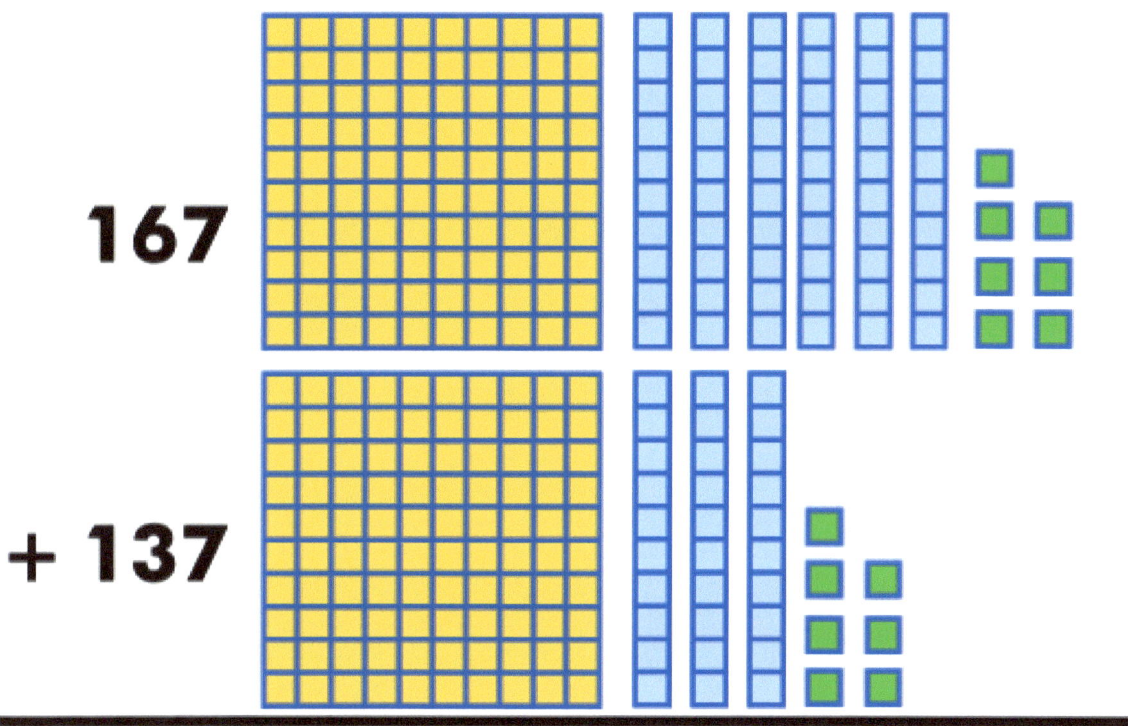

**Adding with Regrouping**

$$4\ 3\ 4$$
$$+\ 3\ 8\ 7$$

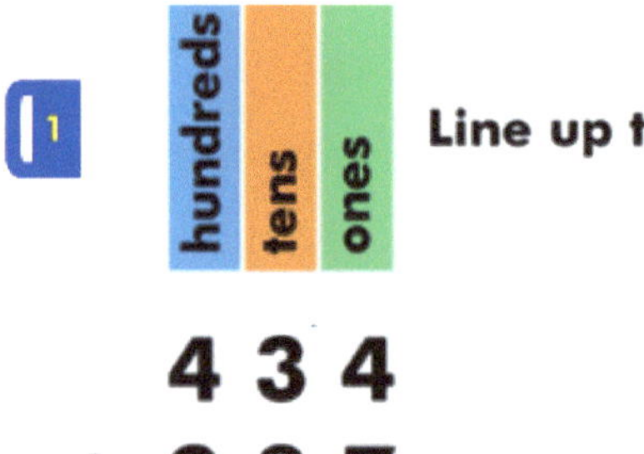


$$4\ 3\ 4$$
$$+\ 3\ 8\ 7$$

**Add the ones:**
4 + 7 = 11

**Regroup: trade
10 ones for 1 ten**

**Add the tens:**
3 + 8 + 1 = 12

**Regroup: trade 10
tens for 1 hundred**

**Add the hundreds:**
4 + 3 + 1 = 8

**Solution: 821**

**Model the solution to this subtraction problem.**

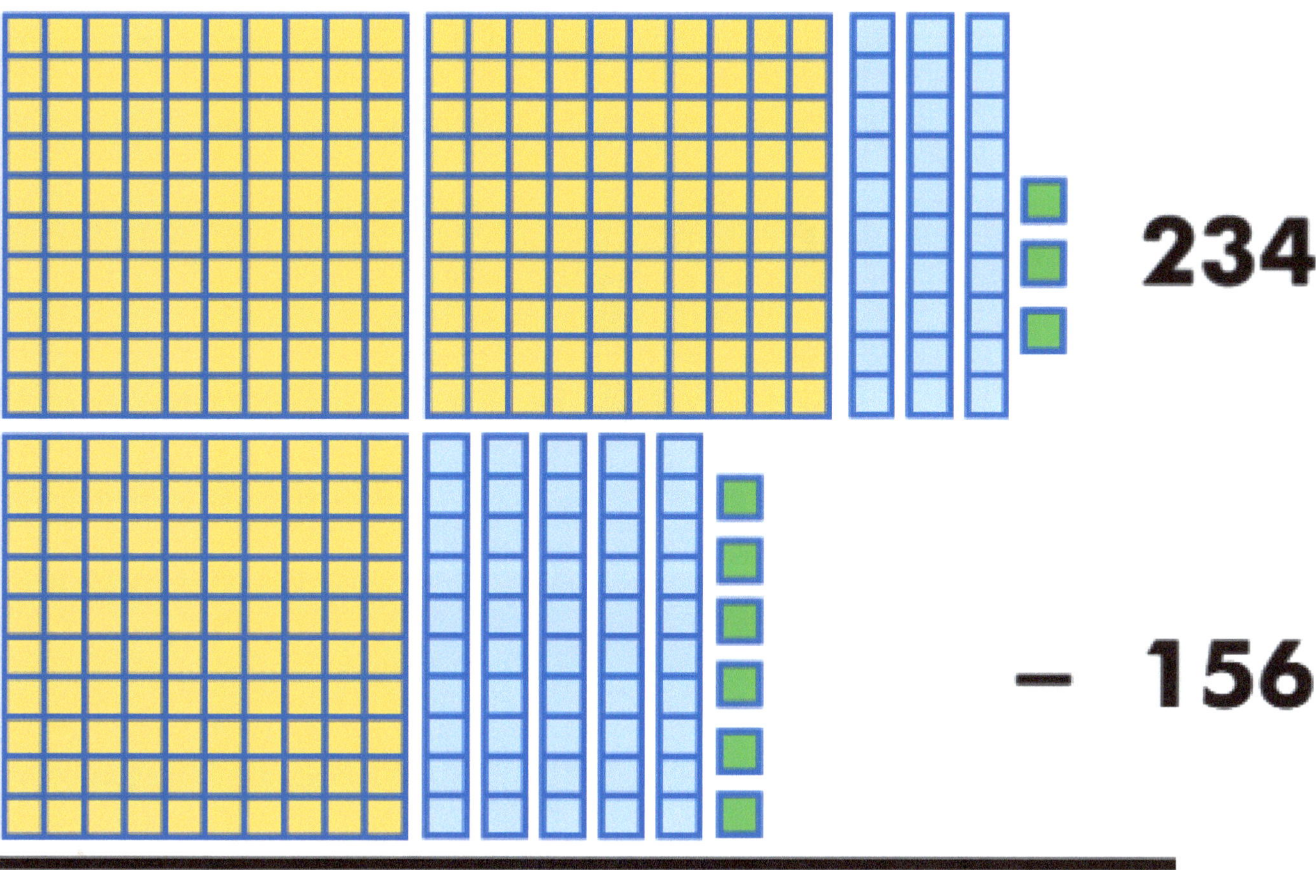

**Subtracting with Regrouping**

$$
\begin{array}{r}
7\ 1\ 3 \\
-\ 3\ 4\ 5 \\
\hline
\end{array}
$$

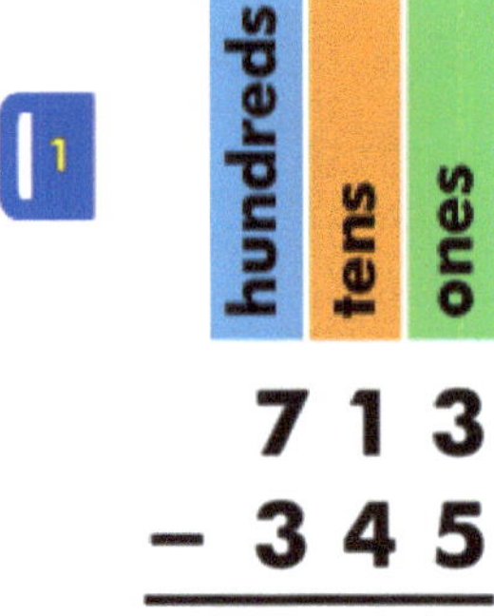


**1**  Line up the place values

$$
\begin{array}{r}
7\ 1\ 3 \\
-\ 3\ 4\ 5 \\
\hline
\end{array}
$$

**2**  **Subtract the ones: regroup from the tens**
$13 - 5 = 8$

**3**  **Subtract the tens: regroup from the hundreds**
$10 - 5 = 5$

**4**  **Subtract the hundreds:**
$6 - 3 = 3$
**Solution: 368**

**Practice addition and subtraction.**

**(1)**

$$
\begin{array}{r}
373 \\
+\ 248 \\
\hline
\end{array}
$$

**(2)**

$$
\begin{array}{r}
621 \\
-\ 234 \\
\hline
\end{array}
$$

**(3)**

$$
\begin{array}{r}
503 \\
-\ 117 \\
\hline
\end{array}
$$

**(4)**

$$
\begin{array}{r}
625 \\
+\ 789 \\
\hline
\end{array}
$$

**Can you solve these word problems?**

During a school-wide reading contest, the fourth graders read 556 books. The fifth graders read 478 books.
How many books did the two grades read altogether?

How many more books did the fourth graders read compared to the fifth graders?

Which operation should you use for each problem?

Name_________________________________________

## Addition and Subtraction Quiz

**1**   **True or false? To find the sum of two numbers, you subtract them.**

**2**   **312 − 197 = ?**

- **A**   117
- **B**   285
- **C**   115
- **D**   125

**3**   **478 + 593 = ?**

**4**   **Michael has collected 397 baseball cards. Kevin has collected 512 baseball cards. How many more baseball cards has Kevin collected?**

# Multiplication

**Key Vocabulary**

product

factor

partial factor

**If there are seven school buses, and each one of them holds 26 children, what is the total number of students?**

Complete the problem below.  Notice how you can estimate by rounding and then adding single digits and adding a zero at the end.  This will help you to know if your answer is likely to be right or if you only need a rough answer.

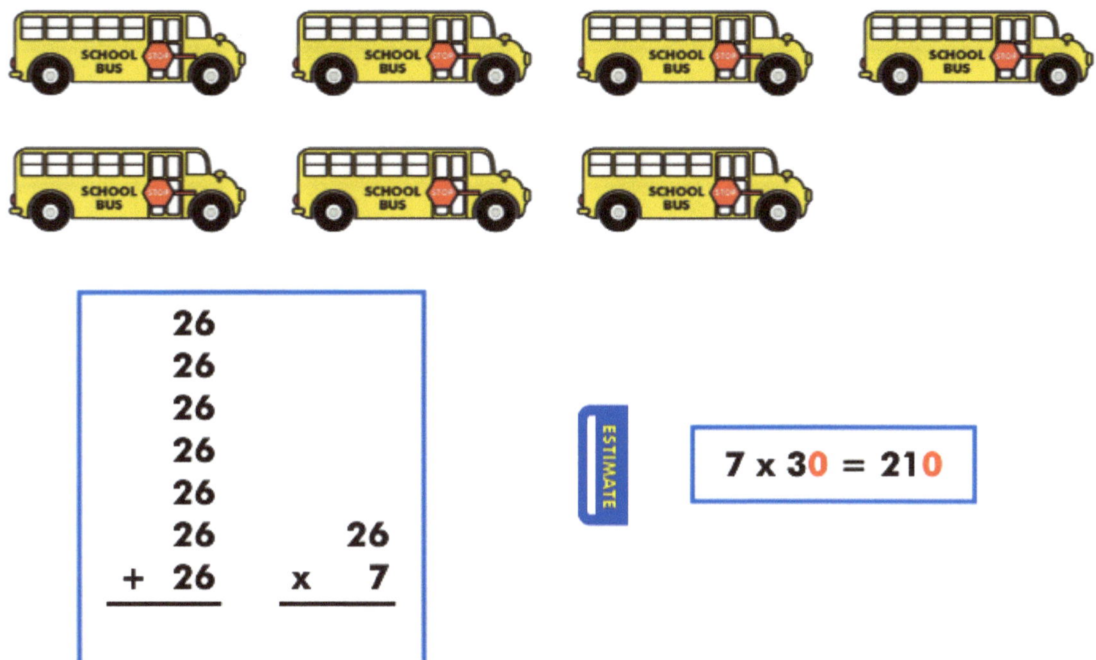

**Multiplying by a One Digit Number**
Sometimes its easier to multiply by a one digit number than two.  In this case we separate 26 into 20 and 6 and then multiply both by 7 (these are called partial products) to get the answer.

Review the illustrations below to discover how we can add partial products to make multiplication easier.

**Partial Products**

**26 x 7 = (20 + 6) x 7**

**= (20 x 7) + (6 x 7)**

**= 140 + 42**

**= 182**

**Area Model**

**Area Model**

**Partial Products.**

Complete the problem below.  The first row has been completed for you to get you started.

$$2\,1\,8 \times 5 = (\boxed{200} + \boxed{10} + \boxed{8}) \times 5$$

$$= (\boxed{\phantom{xxx}} \times 5) + (\boxed{\phantom{xxx}} \times 5) + (\boxed{\phantom{x}} \times 5)$$

$$= \boxed{\phantom{xxxx}} + \boxed{\phantom{xxx}} + \boxed{\phantom{xx}}$$

$$= \boxed{\phantom{xxxxx}}$$

$$\boxed{200 \times 5 = 1{,}000}$$

ESTIMATE

**138 x 5 = ?**

Complete the area model and then calculate the answer using partial products.

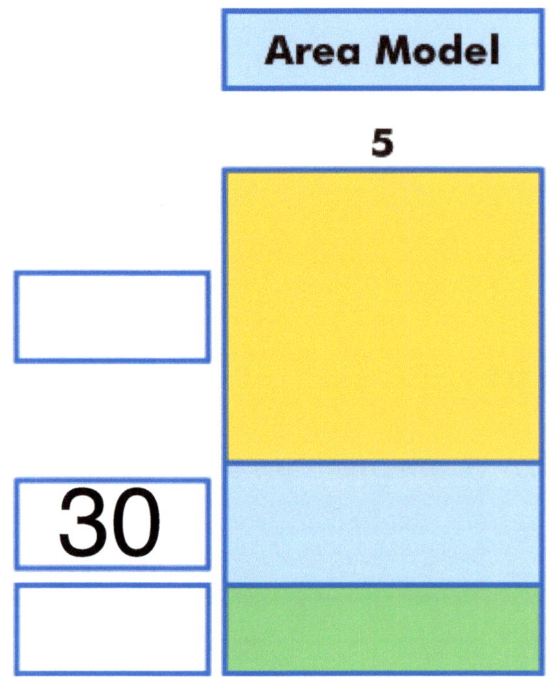

## Multiplying by a Two Digit Number

Review the illustrations below to discover how to use partial products to multiply two digit numbers.  You can estimate to the nearest 10 multiply single digits and add two zeros at the end to make sure you exact answer is close.

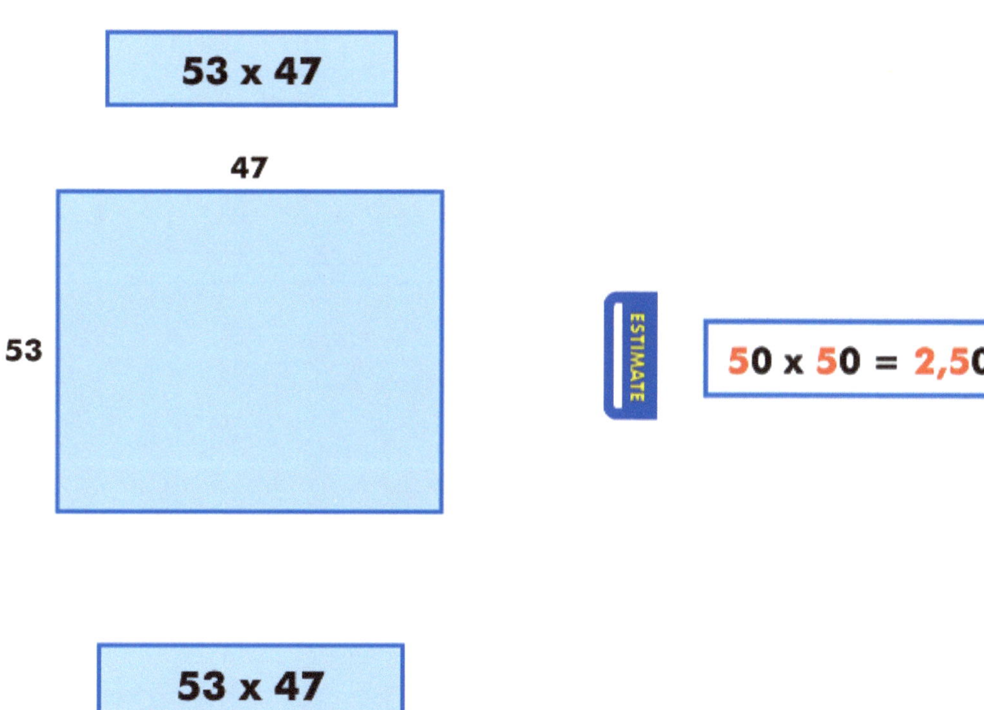


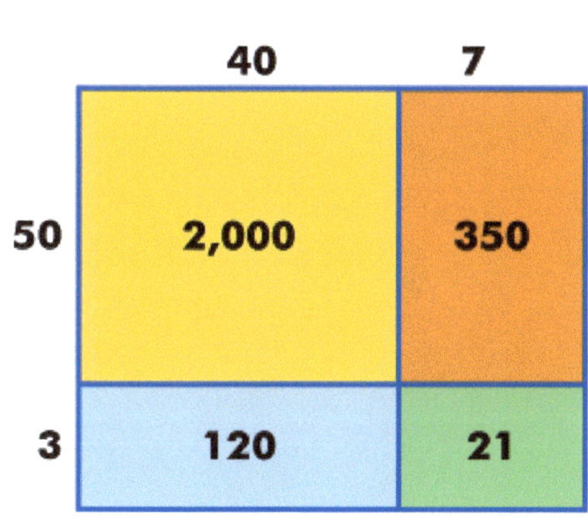


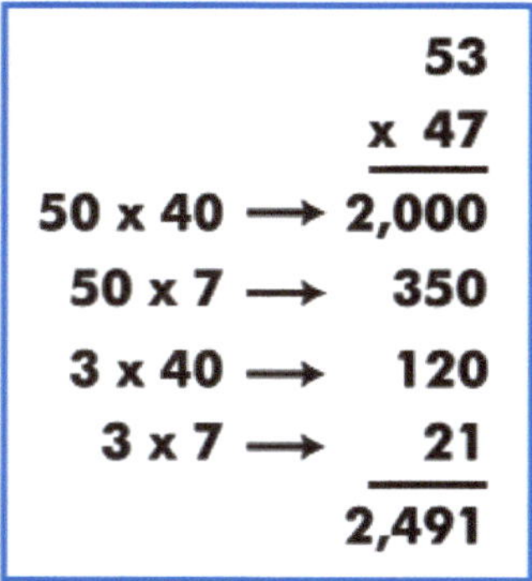

**Practice multiplying by a two digit number**

An estimate has been provided so you can check your answer.  Complete the area model and then calculate the answer using partial products.

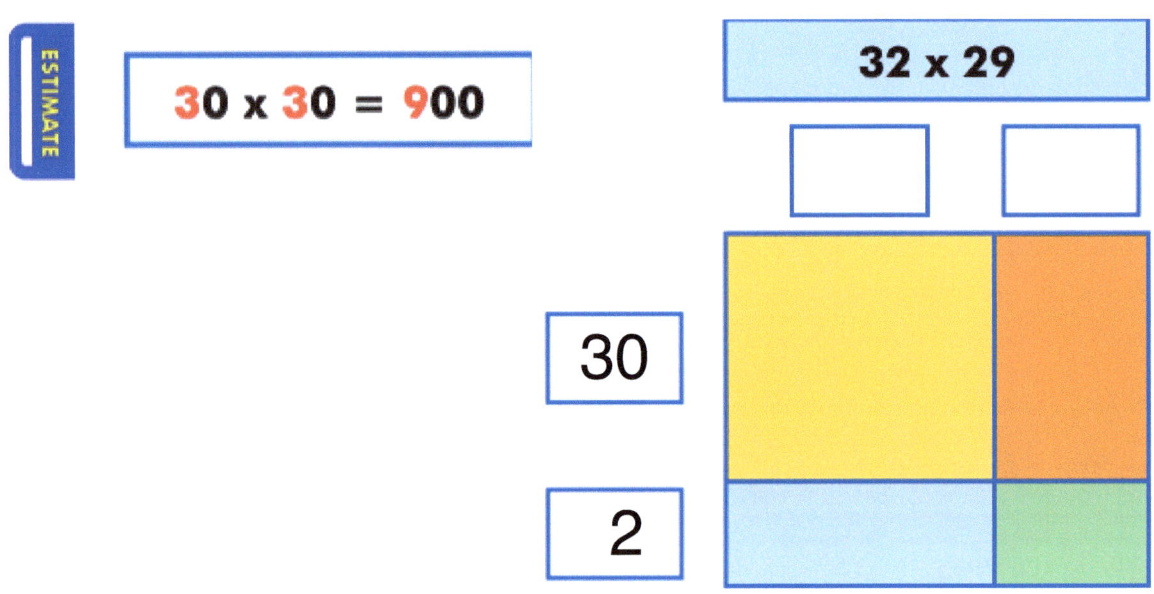

**Apply your new knowledge to a real life problem!**

A marathon is just over 26 miles in length. If the winner of a marathon runs a mile every 5 minutes, and a first-time runner runs a mile every 13 minutes, how many minutes does it take each runner to complete the race? (Use 26 miles for the length of the race).

Estimate for winner is 5 x 25 = ?

Estimate for first-timer is 15 x 25 = ?

Name________________________________________

## Multiplication Quiz

Circle or fill in the correct answer.

 **1**  **22 x 38 = ? True or false, 8,000 is a good estimate.**

 **2**  **43 x 17 = ?**

- **A** 731
- **B** 941
- **C** 321
- **D** 641

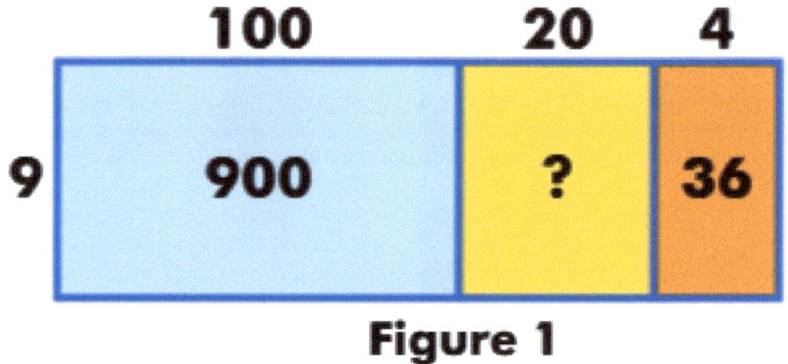


**Figure 1**

**3**  **What is the missing number in Figure 1 above?**

**4**  **124 x 9 = ?**

# Division

**Key Vocabulary**

remainder

divisor

quotient

dividend

**Share $132 between 3 friends.**
Cross off the bills after you add them to a friends box.

$$132 \div 3 =$$

$$\$132 = \$100 + 20 \ + \ \$10 + \ 2$$

To share among three friends, break the bills down further.

$10   $10   $10   $10   $10   $10   $10   $10
$10   $10   $10   $10

$1   $1   $1   $1   $1   $1   $1   $1   $1   $1
$1   $1

**Divide 103 by 7.**
Break down 103 so that you can divide the number more easily.
Discover the process below

$$103 \div 7 =$$

___________

$$70 + 33 = 103$$

$$70 + 28 + 5 = 103$$

___________

$$70 \div 7 = 10$$

$$28 \div 7 = 4$$

5 is left over or is the remainder (R)

___________

$$10 + 4 = 14$$

$$103 \div 7 = 14 \text{ R}5$$

**Your turn!**

$$155 \div 11 =$$

$110 + \rule{2cm}{0.4pt} = 155$

$110 + \rule{2cm}{0.4pt} + 1 = 155$

$\rule{2cm}{0.4pt} \div 11 = \rule{2cm}{0.4pt}$

$\rule{2cm}{0.4pt} \div 11 = \rule{2cm}{0.4pt}$

$\rule{2cm}{0.4pt}$ is the remainder (R)

$\rule{2cm}{0.4pt} + \rule{2cm}{0.4pt} = \rule{2cm}{0.4pt}$

$155 \div 11 = \rule{2cm}{0.4pt}\ R\rule{1.5cm}{0.4pt}$

**Everyone needs a seat.**
See if you can solve a real world problem using your new knowledge.

If 484 students need to be seated in rows that hold 14 students, how many rows will there need to be?

Name_______________________________________

## Division Quiz

**1**  **True or false, a good estimate for 324 ÷ 12 is 30.**

**2**  **324 ÷ 7 ?**

   **A**  **64 R2**

   **B**  **46 R4**

   **C**  **64 R2**

   **D**  **46 R2**

**3**  **Find the remainder:  126 ÷ 10**

**4**  **Find the remainder: 68 ÷ 11**

Newburyport, MA 01950

1-800-596-3175

OnBoard Academics employs teachers to make lessons for teachers!  We create and publish a wide range of aligned lessons in math, science and ELA for use on most EdTech devices including whiteboard, tablets, computers and pdfs for printing.

All of our lessons are aligned to the common core, the Next Generation Science Standards and all state standards.

If you like our products please visit our website for information on individual lessons, teachers licenses, building licenses, district licenses and subscriptions.

Thank you for using OnBoard Academic products.